Angela Royston

www.raintreepublishers.co.uk
Visit our website to find out
more information about
Raintree books.

To order:
☎ Phone 0845 6044371
🖷 Fax +44 (0) 1865 312263
🖳 Email myorders@capstonepub.co.uk

Customers from outside the UK please telephone +44 1865 312262

Raintree is an imprint of Capstone Global Library
Limited, a company incorporated in England and Wales
having its registered office at 7 Pilgrim Street, London,
EC4V 6LB - Registered company number: 6695582

Raintree is a registered trademark of Pearson Education
Limited, under licence to Capstone Global Library
Limited

Text © Capstone Global Library Limited 2009
First published in hardback in 2009
Paperback edition first published in 2010
The moral rights of the proprietor have been asserted.

Edited by Sarah Eason and Leon Gray
Designed by Calcium and Geoff Ward
Original illustrations © Capstone Global Library
Limited 2009
Illustrated by Geoff Ward
Picture research by Maria Joannou
Originated by Dot Gradations Ltd
Printed and bound in China by CTPS

ISBN 978 0 431162 78 2 (hardback)
13 12 11 10 09
10 9 8 7 6 5 4 3 2 1

ISBN 978 0 431162 90 4 (paperback)
14 13 12 11 10
10 9 8 7 6 5 4 3 2 1

British Library Cataloguing in Publication Data
Royston, Angela
Climate change. - (Headline issues)
363.7'3874
A full catalogue record for this book is available from
the British Library.

Acknowledgements
We would like to thank the following for permission
to reproduce photographs:
Corbis: Darrell Gulin 22, Matthieu Paley 17t, John
Sevigny/EPA 9, Mike Theiss/Ultimate Chase 17b;
Dreamstime: Penelope Berger 28, Melvin Lee 21, Kenneth
Roberts 3; Getty Images: Marko Georgiev 11b, National
Geographic/Paul Nicklen 14b, Stone/Arnulf Husmo 24;
Istockphoto: 24–25, Jaap Hart 25, 30–31; Photolibrary:
Robert Harding Travel/Tony Waltham 15r; Rex Features:
Adam Gasson 11t, Richard Jones 26, Peter Oxford/
Nature Picture Library 23, Sipa Press 8, 19; Science
Photo Library: Cape Grim B.A.P.S./Simon Fraser 5t,
Nigel Cattlin 18; Shutterstock: 6, 10, 11, 20, 26–27, Kevin
Britland 12–13, Chrislofoto 29, Jordi Espel 18–19, Gregory
Gerber 13, Chris Howey 22–23, Matsonashvili Mikhail
16–17, Knud Nielsen 4–5, Nik Niklz 21b, Gregory Pelt
8–9, PhotoSmart 1, 12, 13, Svetlana Privezentseva 14, 32,
Adrian Reynolds 27, Hiroyuki Saita 21t, Brad Sauter 7t,
Dan Tataru 5b, Wrangler 7b, Ivars Linards Zolnerovics 7,
Alexandr Zyryanov 15.

Cover photograph reproduced with permission of
Photolibrary/Robert Harding Travel/Tony Waltham.

Every effort has been made to contact copyright holders
of material reproduced in this book. Any omissions will
be rectified in subsequent printings if notice is given to
the publishers.

Disclaimer
All the Internet addresses (URLs) given in this book
were valid at the time of going to press. However, due to
the dynamic nature of the Internet, some addresses may
have changed, or sites may have changed or ceased to
exist since publication. While the author and Publishers
regret any inconvenience this may cause readers, no
responsibility for any such changes can be accepted by
either the author or the Publishers.

Contents

Some words are printed in bold, **like this**. You can find out what they mean by looking in the glossary on page 30.

Climate: Expect the unexpected

THE WEATHER SEEMS to be going crazy. The **climate** is changing in many places. Climate describes the type of weather a place can expect to have at any given time. In many parts of the world, months that are normally cool have become as hot as summer. People do not know what kind of weather to expect. Storms, floods, **droughts**, and other **extreme weather events** are becoming more common.

Extreme weather

We call storms that occur in **tropical** areas **hurricanes, typhoons**, and **cyclones**. These storms are becoming more severe. Their winds are stronger and cause more damage. They whip up huge waves that flood the land. Heavy rainstorms cause flooding, too.

On the other hand, some places have very little rain for many years. They are suffering from severe drought. Crops are failing and the people who live there are starving.

Planet heating up

Extreme weather events and unpredictable climate are due to **global warming**. Global warming is a small increase in the average **temperature** at the Earth's surface. The actual rise in average temperature is less than 1°C (1.6°F), but it is having a huge effect. It is causing climates to change around the world.

Australia, Spain, and many other countries are becoming hotter and drier than they used to be. Many other changes are happening too, including more droughts and more floods. Many scientists prefer to use climate change instead of global warming to describe all these problems.

FACT!

If the Earth's temperature goes on rising, these and other disasters could occur:
✦ More places will have severe droughts
✦ Millions of people will starve as farm crops die from lack of water

BEHIND THE HEADLINES

Meteorologists on the front line

Meteorologists are scientists who study the weather. They measure and record the weather at thousands of places around the world. For example, they record the temperature of the air every day, year after year. Meteorologists tell us how climates are changing around the world. They try to **predict** how they will change in the future.

A meteorologist measures the air temperature at a weather station in Tasmania, Australia. Meteorologists also measure wind speed and how much rain falls every day.

A bolt of lightning strikes a building during a storm in Chicago, USA. Violent storms are becoming more common due to the effects of global warming.

It's official – it's our fault

Climate change has happened before, but it has never happened as fast as it is happening now. This is because people are making it much worse.

Global warming

Coal, **oil**, and natural gas are called **fossil fuels**. When people burn fossil fuels, they make **carbon dioxide** and other gases. The gases collect in the Earth's atmosphere and trap some of the heat from the Sun. As more heat is trapped, the Earth is gradually getting warmer. This is **global warming**. The solution to the problem of global warming is to stop burning fossil fuels, but that is much easier said than done.

Burning fossil fuels

People burn fossil fuels for two reasons. Oil is fuel for aeroplanes, buses, cars, and other vehicles. Burning oil as a fuel drives the engines of these vehicles. Fossil fuels are also burned in **power stations** to **generate** electricity. People use electricity to light their homes and power their computers and other electrical machines.

People in rich countries own more cars and fly in aeroplanes more often than they used to. People in Brazil, China, and India are also burning more fossil fuels than they once did. We have to change the way we do things to slow down global warming.

Are there alternatives?

Some power stations do not burn fossil fuels, but there are not enough of them to produce all the electricity we need. We already have other ways of powering cars, but these cars are not yet cheap enough for everyone to buy.

FACT!

Oil is used to make many important things, including:
✦ medicine ✦ paint ✦ plastic ✦ shampoo ✦ washing-up liquid

Coal arrives at a power station in railway wagons. Coal and other fossil fuels formed from the remains of plants and animals that lived millions of years ago.

BEHIND THE HEADLINES

Is it true?

Some people say that global warming is not our fault. They point out that in the past the Earth has been warmer and cooler than it is now. They say that there is no need to stop burning fossil fuels. Most scientists think that burning fossil fuels is making the problem much worse. The Earth's temperature has never changed as fast as it is changing now.

The Sun's rays heat the Earth, but much of this heat escapes into space, especially at night. Greenhouse gases in the air stop some of the heat from escaping, making the planet warmer.

Burning petrol produces exhaust fumes that contribute to global warming.

heat trapped by greenhouse gases

escaping heat

gases in the Earth's atmosphere

Cyclone kills thousands

On 2 May 2008, a devastating **cyclone** hit the coast of Myanmar in South-east Asia. It killed at least 76,000 people. This was the second severe storm in the Indian Ocean in six months. In November 2007, Cyclone Sidr killed 3,000 people in Bangladesh. The winds from Cyclone Sidr were stronger than those in Myanmar. More people also died in Myanmar because they did not know the storm was coming.

What causes cyclones?

Cyclones begin over warm tropical seas and oceans. Strong winds start to rotate around a central point. The winds whip up enormous waves as the storm moves across the ocean. When it hits land, the wind and waves batter the coast. **Global warming** makes the seas warmer, so cyclones are becoming more severe and happen more often.

Wind and waves

In Myanmar, more than two million people lost their homes. The winds ripped off roofs and blew down walls. The sea flooded towns and villages along the coast. **Mangrove** trees used to grow along the coast of Myanmar. They protected the coast from the waves. However, the mangrove trees had been cut down to clear the land for farming. This made the flooding much worse.

These people lost their homes when the cyclone hit Myanmar in May 2008.

ON THE SPOT
Hurricane Wilma

In October 2005, one of the strongest recorded **hurricanes** hit the Caribbean, Mexico, and Florida in the United States. People knew that the storm was coming and many travelled inland to escape it.

Others sheltered in schools and other strong buildings. The wind and rain flattened thousands of homes. Wilma killed at least 22 people and caused billions of dollars worth of damage.

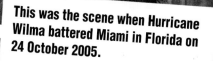

This was the scene when Hurricane Wilma battered Miami in Florida on 24 October 2005.

Flooding devastates New Orleans

On 31 August 2005, the American city of New Orleans in Louisiana was flooded. Masses of water started to collect in Lake Pontchartrain when **Hurricane** Katrina hit land on 24 August. New Orleans was protected by walls called **levees**. These were supposed to keep the water out. However, the walls were not strong enough. They burst in several places, and water poured into the city.

Flood damage

The floods covered most of New Orleans. Streets became rivers and houses were filled with water. Nearly 2,000 people lost their lives. Eventually, the water subsided, and the levees were repaired. The damage to houses and buildings has taken much longer to repair. Many families never returned to New Orleans. People were angry. The levees had failed because the **city council** had not kept them in good repair.

Causes of floods

Flooding along the coast is usually caused by severe storms such as Hurricane Katrina. Inland flooding is usually caused when rivers burst their banks. This is most likely to happen when it has rained heavily for many days. Inland flooding also occurs in spring when winter snow suddenly melts and pours into streams and rivers. **Climate** change is adding to the problem. It results in more severe storms, which causes more floods.

Police officers rescue a man from flood waters that devastated New Orleans in August 2005.

ON THE SPOT
River Severn, England

Flood water from the River Severn covers fields near the town of Tewkesbury.

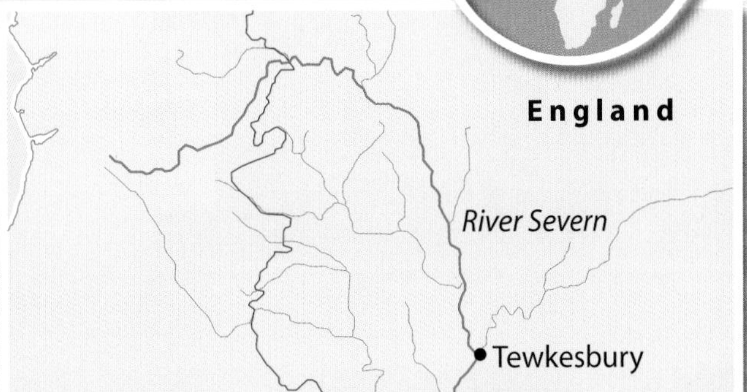

In July 2007, the River Severn in England burst its banks. Water flooded fields and towns. It also threatened to flood the **power station** at Walham in Gloucestershire. The power station would have been closed down if water had got into it. This would have left half a million people without electricity. Fire crews worked through the night to pump out the water. They built a barrier around the station to keep the flood waters out.

England

River Severn

• Tewkesbury
• Walham

Wales

Dying of thirst

IN 2007, SOUTHERN Europe was scorched by a **heat wave**. Forest fires spread across the land, burning vast areas. At the same time, much of Australia, the United States, and southern Africa were suffering from their worst-ever **drought**. Crops were so short of water that they died.

One of the worst affected areas was the Murray-Darling Basin in southern Australia. The drought there had already lasted seven years. Most of Australia's food, including wheat and rice, is grown in the Murray-Darling Basin.

Rising food prices

Australia sells much of the wheat and rice it grows to other countries. As the crops failed in Australia and other countries, the price of these grains began to rise around the world. Wheat and rice are staple foods. This means that they form the largest part of most people's diets. By early 2008, poor people in many countries could not afford to buy bread or rice. Riots broke out in Haiti, Somalia, India, and many other countries.

These maize plants have withered and died because of a severe drought.

◆ The Sahara Desert covers most of North Africa and is the largest desert in the world.
◆ Deserts around the world are becoming larger.
◆ In the next 100 years, the area of land suffering from drought could increase by 30 times.

BEHIND THE HEADLINES

More droughts to come

Climate scientists **predict** that droughts will get worse. By 2050, they think that the average amount of rainfall in countries in Africa, Europe, and the Americas may fall by about 20 per cent. This means that many more countries will suffer from droughts. Australians and other people are already getting used to saving water. Much more will need to be done as climate change continues.

The 1,890 square kilometre (730 square mile) Lake Okeechobee in Florida almost dried up during the drought of 2007.

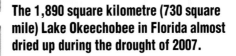

13

Ice caps are melting fast

THE SHEETS OF ice that cover the polar regions are beginning to melt. The **temperature** is rising faster there than elsewhere in the world. Chunks of ice are breaking off and falling into the sea. Melt water beneath the **glaciers** makes them slide faster downhill, too. The ice caps are melting so quickly that sea levels are beginning to rise.

Changing climates

As the ice caps melt, huge amounts of freshwater pour into the salty oceans. This is the reason why sea levels are rising worldwide. Scientists also worry that the extra freshwater in the oceans is causing **climate** change in some countries. They think it is interfering with the North Atlantic Drift.

The North Atlantic Drift is a current of warm water. It flows from the Caribbean Sea across the Atlantic Ocean, past western Europe, and on towards the Arctic. The North Atlantic Drift brings mild weather to western Europe.

Scientists believe that freshwater from the melting ice could stop the North Atlantic Drift from flowing. If this happens, western Europe will have much colder winters than it does now.

The ice that covers the Arctic Ocean is melting. This is making it easier for ships to sail around the north of Canada and Russia.

ON THE SPOT
Himalayas

The Gangotri Glacier in the Himalayas is getting shorter every year. The glacier is part of the thick ice that covers the tops of the high mountains. In the spring, some of the ice from the glacier melts. Water pours downhill providing drinking water and water for crops. Global warming is beginning to melt more of the Gangotri Glacier and many others like it. The extra water is causing **landslides** in the mountains and floods in the valleys and **plains** below.

The Gangotri Glacier is the longest glacier in the Himalayas in northern India. This glacier is getting shorter. It used to reach much farther down the mountainside than it does today.

Islands disappear beneath the waves

THE LEVEL OF the world's seas and oceans is slowly rising. People living on the Carteret Islands off Papua New Guinea are in danger of losing their homes. The rising sea level is threatening to flood the islands, forcing the inhabitants to move elsewhere.

Scientists **predict** that the average sea level around the world will rise by 50 centimetres (20 inches) by 2100. If nothing is done about **global warming**, the sea level could rise by 1 metre (3 feet) or more. Rising sea levels will affect the lives of millions of people around the world.

People at risk

Much of the land in Bangladesh is just 1 metre (3 feet) above sea level. By 2100, 13 million people may have lost their homes. Farmers are losing their land, so they cannot grow crops to support their families.

Some Bangladeshis have already moved inland, but there is not enough space for everyone. Some people who live along the coast will have to ask other countries to take them in.

The causes

The level of the world's seas and oceans is rising for two reasons. As global warming increases the average surface **temperature** of the Earth, the water in the world's seas and oceans becomes warmer. Warm water **expands**. This means that warm water takes up more space than cold water. As a result, the sea level rises.

The second reason for rising sea levels is that global warming is melting **glaciers** around the world. The melt water from these glaciers is also adding extra water to the seas and oceans.

FACT!

Some of the places most likely to be flooded:
- ✦ Kiribati Islands, South Pacific
- ✦ Maldive Islands, Indian Ocean
- ✦ East Anglia, Britain
- ✦ Florida Keys, United States
- ✦ Netherlands
- ✦ Bangladesh

ON THE SPOT
Tuvalu

Nine **coral islands** make up the country of Tuvalu in the South Pacific. Most of the land is less than 1 metre (3 feet) high. The highest point is only 4.6 metres (14 feet). The sea floods onto Tuvalu many times every year. During storms, the waves batter many of the buildings. Scientists think that the 11,000 people who live on these islands will be forced to abandon their country. They will become **climate refugees**.

The islands of Tuvalu in the South Pacific are flat. If the sea level rises any more, the islands will be flooded.

Many houses on the coast of Florida in the United States are in danger of being swept away because of the rising sea level caused by global warming.

Dangerous journey

MANY EUROPEANS GO to the Canary Islands for their holidays. However, tourists are shocked when small boats carrying weak and dying people arrive on the beaches. These people have sailed all the way from West Africa. Many people die on the way. The Canary Islands belong to Spain. Many people are desperate to live in Europe. This is why they risk such a dangerous journey.

People on the move

The people who travel from Africa to Europe are looking for work. Back home, they cannot feed their families. The land in West Africa is too dry to grow much food. Large fishing boats from rich countries catch all the fish in their waters. They leave very few fish for the local people.

Disasters predicted

Climate change is making it difficult for people to live in West Africa and many other parts of the world. Unless people do something about climate change, millions of people will become **climate refugees**. Millions more will die from **starvation** and disease.

Food will become scarcer as droughts become more common. Many people in Africa already die from a disease called **malaria**. As the Earth becomes warmer, malaria will spread to many other countries.

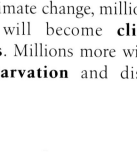

Climate change has turned much of Senagal, in West Africa, into desert. Crops are often planted in poor-quality soil, which makes them very difficult to grow.

Climate refugees should have the right to live in other countries: Who is right and who is wrong?

FOR

Poor countries are worst affected by climate change. The people who live there produce very little **carbon dioxide**. People who live in rich countries are responsible for climate change. They should take in people who have nowhere to live.

A boatload of migrants arrive at the Canary Islands. In 2006, 40,000 people reached the Canaries from West Africa.

AGAINST

Millions of people will become climate refugees. Many countries already take in lots of people. They do not have any more room. Instead they should give poor countries money to help them cope.

Animals doomed to extinction

GLOBAL WARMING AND **climate** change are threatening wildlife around the world. Different kinds of animals and plants live in grasslands, marshes, the oceans, **rainforests**, and many other **habitats**. As the Earth warms up, these habitats will change.

Climate change could make rainforests drier for part of the year. Fires could then destroy large areas of rainforest. The trees of the rainforests take in millions of tonnes of **carbon dioxide** every year. This helps to reduce the impact of global warming. Without these trees, climate change would be even worse. When habitats change, many animals and plants that live in them will die out.

Threats to survival

Many animals are already at risk because of climate change. Polar bears are one example. They live on the edge of the pack ice in the Arctic Ocean. Pack ice is the large blocks of ice that "pack" together on the surface of the ocean. Polar bears hunt seals from this ice.

As more pack ice melts for more months during the year, polar bears could become **extinct**. This means that there will be no polar bears left anywhere in the world. The oceans are becoming warmer. If the sea becomes too warm, the tiny animals that form coral reefs die. Turtles live in the sea, too. They could become extinct if the **temperature** of the sea rises by just 1°C (1.6°F).

Species at risk

A particular kind of animal or plant is called a **species**. Many different species are now in danger of extinction. This is happening faster than ever before.

There are many reasons for this rapid growth in extinction. Scientists think that climate change will make things much worse. If the average temperature of the Earth's surface warms by another 2°C (3.2°F), scientists expect one million species to become extinct. If the Earth warms by 5°C (8°F), four out of every five species could become extinct.

BEHIND THE HEADLINES
Predicting the future

Predicting the future is difficult. Scientists look at what has happened in the past. They set up computer models that tell them what might happen. No one knows for sure whether the predictions will come true. The situation could be better than the scientists predict, or it could be much worse. Some people call this "climate chaos".

Sea temperature affects whether a developing turtle will be male or female. As the sea becomes warmer, fewer male turtles will form. Turtles may die out if there are not enough males to mate with the females.

As the Arctic pack ice melts faster, the polar bears' hunting season may become so short that they will not catch enough seals to survive.

Running out of time

THERE IS STILL time to do something about **climate** change. We can all change the way we do things, but we need to hurry up. As the Earth warms up, certain things will start to happen over which people will have no control. These are called **tipping points**, and they will make the Earth warm up even faster.

Floating ice helps to keep the Earth cool because it reflects the Sun's heat back into space. When the ice melts, the dark water absorbs the heat, adding to global warming.

Arctic ice

In winter, the surface of the Arctic Ocean is completely frozen. White ice reflects light and heat from the Sun. However, in the summer some of the ice melts to reveal the dark water below. This dark water absorbs the Sun's heat. As it does so, the ocean becomes warmer. This melts even more ice, which adds to the problem. Scientists think that all of the sea ice will melt in summer by 2050. This will speed up **global warming**.

Melting tundra

The land around the Arctic Ocean is called the **tundra**. In the winter, the tundra is covered with a thick layer of ice. In the summer, the ice on the surface melts, and the tundra turns into a vast swamp. The ground below remains frozen. This is called **permafrost**. Billions of tonnes of **carbon dioxide** and a gas called **methane** are locked up in the permafrost. As global warming increases the **temperature** of the Earth's surface, more of the permafrost will melt. The gases will then escape into the air, making the Earth warm up even faster.

BEHIND THE HEADLINES
The methane trap

Methane traps more of the Sun's heat than carbon dioxide. Methane forms when the remains of dead plants and animals decay. The tundra is so cold that the moss that grows there only partly rots. If the temperature rises, billions of tonnes of half-rotten moss will decay completely. This will produce huge amounts of methane and carbon dioxide.

In winter the cold tundra is covered with ice. The ice melts in spring and then birds and caribou arrive. They spend the summer there.

Pay now or later?

GOVERNMENTS AND PEOPLE need to prepare now for the effects of **global warming**. When **Hurricane** Katrina flooded New Orleans in 2005, it caused $60 billion of damage. If the **levees** had been kept in good repair, the floods would not have happened.

Start the slow down

We need to prepare for global warming, but we also need to slow it down. The only sure way to do that is to stop burning **fossil fuels**. It is already possible to **generate** electricity without burning fossil fuels. **Hydroelectric power stations** use running water to generate electricity. It is also possible to generate electricity from the power of the wind and heat from the Sun.

We also need to use less energy. People can save energy by installing double-glazing and insulating their homes. Scientists are designing cars that do not burn petrol, but run on electricity. Engineers are also designing new engines that run on **hydrogen**.

Preparing for disaster

Some people think that it is cheaper to adapt to warmer **temperatures**. For example, **city councils** could build more flood walls to protect towns and cities from flooding. People can adapt to **drought** by using less water in their homes and gardens. However, farmers cannot produce food without water. They cannot easily adapt to drought.

Norway generates most of its electricity at hydroelectric power stations such as this one in Alta.

We should adapt to climate change, not try to prevent it: Who is right and who is wrong?

FOR

It is too late to stop **climate** change – it is happening already. It is better to spend money now to prepare for it.

This wall in the Netherlands is called a dyke. It is made of earth and stops the sea from flooding onto the low-lying land. Can it cope with rising sea levels and more storms?

AGAINST

Climate change is happening too quickly. Poor countries do not have time to adapt. People must stop burning fossil fuels. It will cost less money than dealing with the consequences of climate change.

Governments must act

CLIMATE CHANGE AFFECTS all countries. Governments must work together to slow down **global warming**. Rich countries produce the most **carbon dioxide**. They also have enough money to develop cleaner ways of doing things. This will only work if all countries agree to cut the amount of carbon dioxide they produce.

This power station in China burns coal to generate electricity. It is adding to global warming.

China and India

In the past, people in China and India produced very little carbon dioxide. Today, China and India are becoming more industrialized. They are building new **power stations** that burn **fossil fuels**. They are building new factories to make clothes, computers, and other goods. All these new industries are speeding up global warming. People in China and India want to live the same kind of lives as people in Britain, the United States, and other rich countries.

What's the solution?

Governments must agree on how much carbon dioxide each country can produce. The amounts must be low enough to slow down global warming. Scientists must also invent new ways of doing things. These new ideas must be cheap enough for everyone to afford. They must produce little or no carbon dioxide. Much of the technology already exists to cut carbon dioxide emissions. Governments need to invest more to make a difference.

People, not governments, must cut carbon dioxide: Who is right and who is wrong?

FOR

People produce carbon dioxide every time they drive in a car or switch on the television. It is up to each person to cut down the amount of carbon dioxide they produce.

Flying produces more carbon dioxide than any other way of travelling. People should fly only when they really have to.

AGAINST

People cannot stop using their cars unless they take buses or trains, instead. Governments should make bus and train services much better to attract drivers. They should also build power stations that do not burn fossil fuels.

Get involved!

THERE ARE MANY things you can do to help slow down **global warming** and its effects. One of the easiest ways is to save electricity. Make sure you shut down your computer when you are not using it and turn off the television at the mains supply.

Sometimes you cannot avoid travelling by car, but other times you could perhaps take a bus or train, instead. Walking or cycling is even better. The best way to help slow down global warming is to persuade your family and friends to do something about it, too.

Walk to school or take a bus instead of getting a lift in a car. This will slow down global warming and make you fitter.

THINGS TO DO

Travelling

- Avoid travelling by car or aeroplane unless there is no other way.
- For short distances, walk or cycle.
- Use buses or trains to travel longer distances if possible.

If you have a garden, ask your parents to buy a water butt. It will collect rainwater to use in your garden.

Saving electricity

- Don't leave televisions, computers, and DVD players on stand-by mode.
- Don't leave phone chargers plugged into electric sockets.
- Don't leave lights on in empty rooms.
- Wear warm clothes when it is cold and turn down the heating a little.
- Shade the room, open windows, and wear light clothes in hot weather.

Saving water

- Shower instead of bathing.
- Use a cup of water to clean your teeth, instead of a running tap.
- Use a water butt to catch rainwater for the garden.

Investigate climate change

Ask your parents about the **climate** when they were younger.

- Was the weather more predictable then?
- Are the winters colder or milder than in the past?
- Are the summers hotter than they once were?
- Did they see birds, insects, or other animals that you don't see today?
- Do you see wild animals now that they didn't see before?

Remember that things change for many different reasons. Global warming is just one possible reason for a particular change.

Glossary

carbon dioxide one of the gases in the air. Carbon dioxide traps the Sun's heat and so leads to global warming.

city council group of people who are usually elected to govern a city

climate types of weather that usually occur in a place at different times of the year

climate refugee person forced to abandon his or her home because of changes due to global warming

coral island island that is formed by small animals called coral polyps. The polyps build their shells on top of previous polyps. Over millions of years, the coral polyps form a reef or an island.

cyclone rotating storm with fierce winds and heavy rain

drought long period without rain

expand increase in volume

extinct when no member of a species still exists in nature

extreme weather event very severe weather, such as a blizzard, hurricane, thunderstorm, or heavy rainstorm

fossil fuel fuel such as oil and gas formed from the remains of plants and animals that lived millions of years ago

generate produce

glacier area of thick ice high on a mountain or in the Arctic or Antarctica. Glaciers slide very slowly downhill.

global warming increase in the average temperature at the Earth's surface

habitat kind of place where animals and plants normally live in the wild

heat wave when the weather is hotter than usual for several days or longer

hurricane severe storm that begins in the Atlantic Ocean or eastern Pacific Ocean

hydroelectric power station system that uses running water to generate electricity

hydrogen gas used as a clean fuel because it produces water when it burns.

landslide when part of a hillside breaks away and slides downhill

levee raised bank or wall built along rivers or lakes to stop water flooding onto the land

malaria disease caused by germs that are spread by some types of mosquitoes

mangrove tree that grows in salty water along the edge of the coast. Mangroves help to protect the land from the sea.

meteorologist scientist who studies and predicts the weather

methane gas that is produced when living things rot

oil liquid that forms under the ground and is burned as a fuel

permafrost ground that stays permanently frozen in the polar regions

plain flat land, often between mountains and the sea

power station building in which power is created, such as an electrical power station

predict tell how things might change

rainforest thick forest that grows where it rains heavily almost every day

species particular kind of living thing

starvation not having enough food

temperature how hot or cold something is

tipping point something that happens over which people have no control

tropical from part of the Earth near or on the Equator

tundra swampy land around the Arctic Ocean that is frozen in winter

typhoon severe storm. Typhoons are also known as hurricanes.

Find out more

Books

Changing Climate (Earth Watch), Sally Morgan (Sea to Sea Publications, 2007)

Changing Climate: Living with the Weather (Geography Focus), Louise Spilsbury (Raintree, 2006)

Climate Change (Can The Earth Cope?), Richard Spilsbury (Wayland, 2008)

Climate Change (DK Eyewitness Books), John Woodward (DK Publishing, 2008)

Extreme Weather: Science Tackles Global Warming and Climate Change (National Geographic Investigates), Kathleen Simpson (National Geographic Society, 2008)

Websites

Find out about climate change at the website of the United States Environmental Protection Agency:
http://epa.gov/climatechange/kids/index.html

This website tells you all about the world's rainforests, why they are important, and what you can do to help save them:
http://kids.mongabay.com/

Find out about climate change and what you can do to help with Tiki the penguin:
http://tiki.oneworld.net/global_warming/climate8.html

This award-winning website tells you all about climate change and what you can do to help. Click on the "Causes & Effects" link to find out about polar bears, coral reefs, and rising sea levels:
www.coolkidsforacoolclimate.com

Index